JUNKERS JU 52

A restored Ju 52 in flight (1986) over the area of Bremen. (Lufthansa photo archives)

Heinz J. Nowarra

Schiffer Military/Aviation History
Atglen, PA

SOURCES/PHOTOS

- **Lufthansa photographic archives**
- **Morzik,** *Die deutschen Transportflieger*
- **Nowarra,** *Die deutsche Luftrüstung*, **volume 3**
- **Zindel,** *Geschichte und Entwicklung des Junkers-flugzeugbaus 1910-38 und bis zum Ende 1970*
- **Junkers, factory files**
- **Nowarra personal archives**

TANTE JU ON THE BANKS OF SICILY
Despite the dire need of air transports for the impending spring offensive in the Russian campaign, the beleaguered Transportverband was diverted first to the rescue of Wehrmacht troops escaping the collapse of the North African front. In April of 1943, the 12th Air Force Western Air Command initiated Operation Flax, a concerted offensive to eradicate those rescue efforts. In the early morning hours of April 5th, an armada of fifty-plus Junkers transports with heavy fighter escort was intercepted by P-38s just off the west coast of Sicily. Even though victory claims by the Lightning pilots were highly exaggerated, actual losses of the fuel and equipment laden "Tante Ju" (Auntie Ju) transports were staggering. Following Allied fighter sweeps of the Tunisian landing fields and another interception of the return flight to Sicily that evening, twenty-five Ju 52s were officially written off the Luftwaffe records for the day. By month's end, an estimated one-fifth of the Transportverband's tri-motor types would be lost on the Mediterranean front.

Translated from the German by James C. Cable.

This book originally appeared under the title,
Junkers Ju 52,
by Podzun-Pallas Verlag, Friedberg.

Copyright © 1993 by Schiffer Publishing Ltd.

All rights reserved. No part of this work may be reproduced or used in any forms or by any means – graphic, electronic or mechanical, including photocopying or information storage and retrieval systems – without written permission from the copyright holder.

Printed in the United States of America.
ISBN: 0-88740-523-1

We are interested in hearing from authors with book ideas on related topics.

Published by Schiffer Publishing Ltd.
77 Lower Valley Road
Atglen, PA 19310
Please write for a free catalog.
This book may be purchased from the publisher.
Please include $2.95 postage.
Try your bookstore first.

Over Holland in 1940.

Ju 52 ba, serial number 4001, was presented to the public for the first time on February 17th 1931 at Berlin Tempelhof airport in Berlin.

THE JUNKERS JU 52 at WAR AND IN PEACE

The construction of this most famous of German transport aircraft was really the result of a purely commercial consideration by the Deutsche Lufthansa management. It started as early as 1925 when Dipl.Ing. Kurt Weil and Hans M. Bongers asked the "Junkers Luftverkehr" air transport division for a commercial three-engine transport plane. But it was not until 1928, two years after the founding of "Deutsche Lufthansa" that Junkers began to consider taking up the development of a new commercial aircraft. Ernst Zindel, who was charged with the completion of the construction, concentrated on two versions of the same model – now called the J 52, these versions were a single-engine freight aircraft for the air transport company envisioned by Prof. Junkers called "Luftfrako International," and a three-engine aircraft for up to 17 passengers, in keeping with the requirements set by Deutsche Lufthansa (DLH). After the war, Dipl.Ing. Zindel had this to say about both models:

The Ju 52 wing displaying the Junkers external flaps.

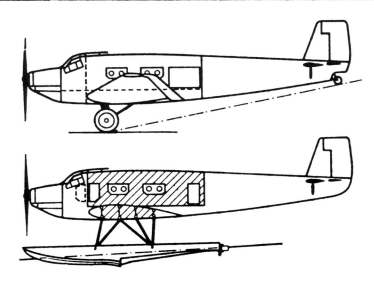

Junkers Ju 52 be
Specifications

Length .... 18.5 meters
Wingspan .... 29.5 meters
Wing area .... 116.0 meters^2

Weights
ramp weight .... approx 3850 kilograms
useful load .... approx 3150 kilograms
takeoff weight .... 7000 kilograms
wing loading .... 60.30 kilograms/m^2

Performance
Max speed, low altitude ...... 194 km/h
Cruise speed, low alt .... 160 km/h
landing speed .... 77 km/h
rate of climb:
  0-1000 meters in .... 9.8 min
  0-2000 meters in .... 22.4 min
  0-3000 meters in .... 42.0 min

Absolute max. altitude .... 4200 meters
Sevice ceiling .... 3200 meters
take-off run .... 255 meters
landing run .... 155 meters

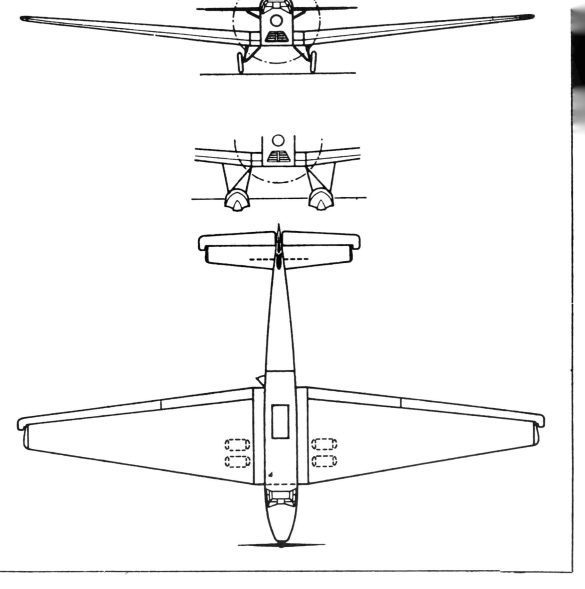

"The single-engine transport was easily recognizable by a large, 1.6-meter wide and 1.5-meter high loading hatch together with a loading ramp on the rear portion of the side of the fuselage, easily accessible behind the wing's trailing edge and external flaps, the lower portion of which served as a loading ramp; further, there was a 1.5m X 1.2m cargo hatch in the bay for loading heavy and bulky freight from above the aircraft's center by means of a crane. For crew entry and exit, there was a door just behind the cockpit. The wing, as was with previous production models, was equipped with external airfoil flaps and ailerons to increase the maximum lift during take-off and landing. With a useful load of 2000kg and a range of about 1200 kilometers, the take-off weight of this single-engine aircraft was about 7000 kilograms. With a wingspan of 29.5 meters and a wing surface area of 110 square meters, wing loading was approximately 64kg/m^2 and the landing speed approximately 82 km/h. Because the aircraft had very good flight characteristics during landing and a robust landing gear, it could take off and land on regular grass strips without problem, which was of great importance for a transport aircraft.

In the case of the passenger aircraft, which was to operate in normal service from prepared airfields with better runways, the take-off weight and wing loading could be substantially higher even with the same wing dimensions: with a maximum of 17 passengers, the necessary furnishings and comfort items (heating, ventilation, heat insulation and additional radio equipment), sufficient luggage and postal freight and a range of 1300 km, its take-off weight was approximately 10,000kg with a wing loading of 91 kg/m^2, which is about 40% more that in the

The Ju 52 ba's main wheel and oleo struts.

The loading hatch in the upper fuselage made loading of even bulky items possible.

Ju 52 di, serial number 4002, in October 1931 after sea trials had been broken off.

Ju 52-3m, serial number 4013, registration number D-2201, was the first three-engine Ju 52 to be delivered to Lufthansa.

case of the transport aircraft, and a correspondingly 20% faster approach speed of about 100km/h.

For the powerplants we selected the best and most reliable engines which could be obtained at the time for both the single-engine transport and the three-engine passenger aircraft. For the single-engine aircraft it was necessary to utilize a high-performance motor, and high-performance motors were still quite rare in 1929. It was well within our desires that at that time two relatively high-performance engines became commercially available:

In Germany, this was the water-cooled BMW VIIa, a twelve-cylinder V engine which had been developed from the well-known and proven BMW IV and BMW VI engines and was the first engine to be equipped with a reduction gear for the propeller shaft, which allowed the propeller's rpm's to be reduced to achieve the best possible performance from the 700 horsepower engine. This engine was to later be replace by the Jumo L88, a liquid-cooled twelve-cylinder engine which was still under development and not quite ready for deployment, which, with its 800 horsepower, provided a 100 horsepower increase in performance during takeoff. This engine was, like the in-line six-cylinder Jumo L8, created by improvements made to increase the RPM speed of the proven Jumo L5 and its doubled L55; its crankshaft could operate at maximum of 2000 rpm; the propeller shaft was reduced to 800 rpm via a reduction gear.

The second engine available for use which corresponded to our power requirements was the air-cooled "Leopard" 14-cylinder twin radial engine manufactured by the British firm of Armstrong-Siddley, which had a take-off power

Serial number 4016, a Ju 52-3m ba, was a special version, which was delivered to the Prince Bibesco, President of the Federation Aeronautique International (FAI).

This Ju 52-3m ci, serial number 4014, was outfitted as a float plane and deliverd to the Finnish "Aero O/Y" airline.

Ju 52-3m ho, serial number 4045, was one of two Lufthansa Ju 52's which were equipped with Jumo 205 diesel engines.

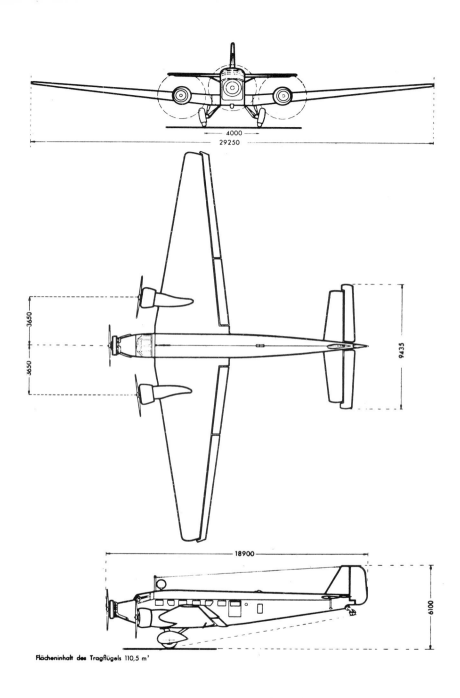

Factory plans for the Ju 52-3m ge, land version.

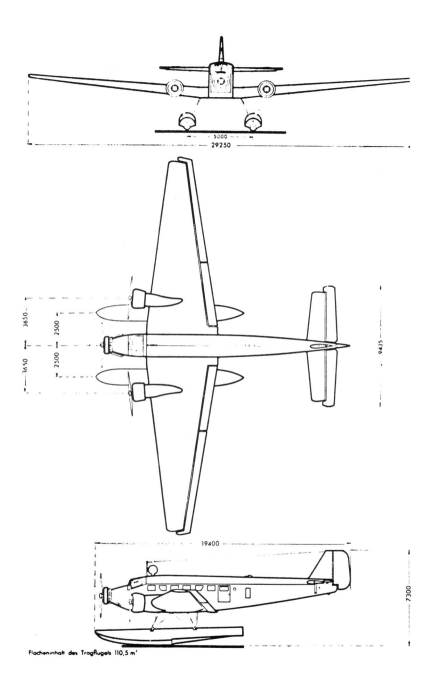

Plans for the Ju 52-3m ge, float plane version.

This Ju 52-3m, SE-AFD, serial number 5646, was one of the seven Ju 52's to be delivered to Sweden. It was later sent to IBERIA in Spain as EC-ADP.

Ju 52-3m ge, serial number 5478, on Rhein-Main airport. This aircraft was later sent to the Condor Syndicate in Brazil as PP-CBD.

of 800hp and was also equipped with reduction gear for the propeller shaft. In regards to sales of aircraft to foreign customers, it was a fine bit of luck that the British Armstrong-Siddley engine was available in addition to the German water-cooled engine.

Despite the fact that the aircraft in use by Canadian Airways were particularly well-suited to the unusual terrain and special needs of the area, the single-engine Ju 52 did not find the sales which were hoped for, particularly undershooting the great plans for air freight commerce envisioned by the old Junkers air traffic experts, simply because the times and the general conditions were not ready for these types of ideas."

The first single-engine Ju 52 ba (serial number 4001) made its maiden flight on October 13th 1930 equipped with a Junkers L-88 engine, which was afterwards changed to a BMW IIau, redesignated the Ju 52 be and was presented to the public for the first time at the Berlin-Tempelhof airport on February 17 1931 with the registration of D-1974. The "Luftfrako" did accept the aircraft, but returned it to Junkers in July. What followed were aircraft serial numbers 4002, 4003 an 4004 with an improved rudder and designated the Ju 52 ce. The first two remained with Junkers, but in 1932 serial number 4004 D-2317 was sent to the Deutsche Verkehrsfliegerschule (DVS) as a glider airplane and later to Sweden as SE-ADM. The following serial number 4005 (D-2356) with BMW IX U engines went to the Reichsverband der Deutschen Flugzeugindustrie (RDL), which was a covername for the secret aviation units of the Reichswehr. They were destroyed by fire in May 1934.

What followed was a Ju 52 di (serial number 4005) with the 750hp Armstrong-Siddley "Leopard" engine – first fitted with landing gear, and then delivered to Canada in 1931 as a floatplane. There it was retro-fitted with a Rolls-Royce "Buzzard" engine and redesignated as Ju 52 cao. This aircraft flew with Canadian markings under the most difficult of conditions until 1947. Serial number 4007 D-UHYF, 4002, first D-2133 and later D-USUS and 4003 D-USON were utilized by the Luftdienst-Schleppstaffel in Kiel-Holtenau as target-towing aircraft for the Reichsmarine. No further single-engine Ju 52 aircraft were built; no one wanted them. On the other hand, there were constant inquiries about the three-engine version. Ernst Zindel wrote about this first three-engine model in his memoirs:

"In the original model destined for passenger service, we decided once again on the three-engine low-wing plane just as with the G 24 and G 31 based on careful consideration and previous experiences. The three-engine model represented the best compromise between flight safety and economy, at least among those aircraft of the size and flight conditions which were being examined.

The external fuselage, the aerodynamic characteristics and the basic layout were practically the same for the single-engine and the three-engine variants, only the stress points and landing gear were reinforced due to the greater weight and flying speeds of the three-engine model.

We selected the air-cooled nine-cylinder Hornet radial engine manufactured by the American Pratt & Whitney company for the three-engine model to be used in international

Ju 52-3m di, serial number 4010, nr. 621, of the Columbian Air Force.

Below: This Ju 52 belonging to the German/Russian airline DERULUFT was fitted with skis in the winter.

IBERIA bought this Ju 52-3m from the Swedish "Flygtjänst" airline as SE-AFB, serial number 5620, and flew it as EC-ADO.

From the outbreak of WW II until operation "Weserübung" the three Ju 52's flew carrying these markings of neutrality.

air transport. These engines were just beginning production at BMW under license, and initially had a takeoff power of 585hp, but was later replaced by an improved BMW 132, which, in the form utilized in the Ju 52, provided 660hp of takeoff power. For delivery of these aircraft overseas to, for example, South African Airways, an improved version of the Pratt & Whitney Hornet engine with greater performance was installed."

At Junkers, they had prepared serial numbers 4008 to 4012 for final assembly, and when urgent orders for three or two engine aircraft, they could simply fall back on those and retrofit them to three-engine models.

Aircraft serial number 4007, now a three-engine Ju 52-3m ce, was ready for flight in 1931.

Serial numbers 4008 and 4009 were sent to Bolivia, while 4010, 4011 and 4012 went to Columbia. But even with these first aircraft, the fate of the Ju 52 could be seen – all five aircraft did not go, as planned, to airline companies, but rather were used as makeshift bombers and military transports. The two bolivian Ju 52's played a vital role in the so-called Gran Chaco war against Paraguay. Two of the columbian aircraft were lost in 1936 in Cabuyaro and Putumayo, while the remaining four (three others followed the initial order) flew until 1943. The last one was still standing at the Bogota airport in 1966.

Lufthansa received serial number 4013, which was designated D-2201 "Boelke" in May 1932 and redesignated D-ADOM. Serial number 4015 (D-2202 "Richthofen") followed in September. This aircraft immediately aroused foreign interest. Finland became the first foreign customer to receive a Ju 52-3m ce with landing gear and floats and the American Pratt & Whitney "Hornet" engines and was registered as OH-ALK "Sampo." A Ju 53-3m ba equipped with Hispano-Suiza engines was built for the president of the Federation Aeronautique Internationale (FAI). The Ju 52-3m fe was an improved Ju 52-3m with "Hornet" engines which was delivered in 1933. The largest series of the Ju 52-3m was the ge model, which was built in variants from g1e to g14e. Next were the numerous special models which were delivered to foreign customers with various engine types such as the Pratt & Whitney "Wasp," the Piaggio PXR and the Bristol "Pegasus."

Special models were both the Ju 52-3m ho's, D-AJYR and D-AQAR, which were equipped with Jumo 205 C diesel engines. They flew with Lufthansa.

Other models to fly with Lufthansa were: Ju 52-3m gle, g2e with BMW 132 A/E's, geX with BMW 132A-3's, reo with BMW 132 Da/Dc's, te with BMW 132 G/L's and the Ju 52-3mZ, Z1 with BMW 132Z-3's and improved cabin furnishings.

One interesting model was the Ju 52-3m g3e, the makeshift bomber of the new Luftwaffe, which formed the basic inventory of the Luftwaffe's combat units.

When Ju 53-3m g3e's appeared for the first time in the skies above the National Socialist Party's rally at Nürnberg in 1935, the impression they made was so convincing that a French journalist noted in his report: "They manoeuvered with the precision of Frederick the Great's Grenadiers." All of these makeshift bombers were later converted back to transport aircraft at Weserflug.

The Ju 52-3m reigned over european air transport until 1939. The Lufthansa fleet, for the most part, consisted of these aircraft. When the war broke out, 62 of these aircraft were given to the Luftwaffe, and by the end of the

Carl-August Frhr. von Gablenz carried out his famous flight over Pamir with this Ju 52-3m te, serial number 5663, tail number D-ANOY.

Below: A "makeshift" bomber in 1935 – a Junkers Ju 52-3m g3e of KG 353.

In 1938, this Ju 52-3m, D-ADWB, serial number 6650, supported the Nanga-Parbat expedition.

This Lufthansa D-2624, serial number 4026, shown here with the markings of the Flug-Eisenbahn (FLEI), was one of the aircraft used to train the first bomber pilots of the new Luftwaffe before it was made public in 1935.
Below: converting the Ju 52-3m g3e to a makeshift bomber was conducted at Weserflug. The aircraft had to be brought from Nordenham to Einswarden in by ferry with the "Otto" as the tugboat.

war, Lufthansa had only three remaining. Twenty-eight others were "buried," meaning by the Luftwaffe.

The Ju 52-3m g3e followed the g4e version, a military transporter, of which Lufthansa received ten aircraft, three went to Switzerland and still exist today, at least of which is completely airworthy. The g4e could be converted for the following purposes: E = box carrier, R = passenger aircraft with 16 seats, H = flying classroom, St = squadron troop transporter, F = paratroop and air assault aircraft. Engines were the BMW 132 A. What followed that the g5e version, a transporter for land and sea deployment. The Ju 52-3m g6e was similar, but on had regular landing gear or that for snow and had the FuG 3aU radio gear. The Ju 52-3m g7e, similar to the g5e, but with the BMW 132 T engines and the Siemens K4ü g8e autopilot was more like the g6e only with a few interior modifications. There was only a very small series of the Ju 52-3m g12e's built wit the BMW 132 L engines. The g14e was also like the g8e, but had reinforced armor. A few of the g4e to g6e versions were converted to mine-sweeping aircraft and were deployed right up to the end of the war. Of those, a few continued to fly after the cease fire in 1945 under Allied control on anti-mine missions.

The Ju 52 received its baptism of fire in 1935 in Spain. At the request of General Franco, Hitler sent 20 Ju 52's to transport the Spanish troops stationed in Morocco, which were to be deployed to fight the republican units in Spain.

As a cover for this operation, the "Hispano-Marokkanische Transport AG (HISMA) was founded. Oberleutnant Rudolf Freiherr von Moreau was leader of this unit. On the republican side, they began to notice the constant flights of Ju 52's commuting between Tetuan and Sevilla and opened fire on them with anti-aircraft artillery, especially from the ships lying off the coast, one of which was the armored cruiser "Jaime I." In response, two of the Ju 52's were loaded with bombs. Moreau flew one, the other was flown by Flugkapitän Henke of Lufthansa, who had Leutnant Graf Hoyos on board as a bombardier. On 13 August 1935 the two aircraft took off. Moreau missed the target; Hoyos hit the "Jaime I" twice. The hits were so effective that the "Jaime I" had to be towed to Cartagena. From that point on, the transport flights were not disturbed.

In November of 1936, "K 88," the bomber group of the Condor Legion, the Luftwaffe unit supporting Franco, was formed from the Moreau squadron and three other Ju 52-3m g3e's. Moreau created the first national Spanish bomber squadron from 10 Ju 52-3m g3e's. Then came single-seat fighters for the Soviet Union, the Polikarpov I-15. By November 4th 1936, the first Ju 52 piloted by Leutnant Kolbitz was shot down near Madrid by an I-15. On 15 November the squadron had to tangle with 25 I-15's. Afterwards, the Ju 52's were withdrawn from Spain and replaced with Do 17's, He 111's and Ju 86's. From that point on, they served only as transport aircraft.

In the mean time, the Ju 52m had made great strides in international air commerce. The Ju 52 was flying in Austria, Sweden, Italy, South Africa, Mongolia, Mozambique Belgium, Denmark, Norway and Brazil. One very special feat was the flight of a Ju 52-3m, D-ANOY "Rudolf von Thüna" over Pamir to China under the leadership of Freiherr von Gablenz with Flugkapitän Unticht and Oberfunkermaschinist Kirchhoff. They departed from Tempelhof in Berlin on 14 August 1937, but did not return home until October 3rd and only after much difficulty.

Converting the makeshift bombers back into transport aircraft was also conducted at Weserflug. Below: Ju 53-3m g3e of 1./KG 152 en route to the National Socialist's Reichsparteitag in Nürnberg, 1935.

Ju 52-3m g3e: pilot seats and instrument panel.

In the mean time, the first transport unit of the Luftwaffe was created as the "Kampfgeschwader zur besonderen Verwendung (KGzbV) 1." It was created from a branch of the IV./KG 152 "Hindenburg," which was the unit which flew in the 1935 Nürnberg rally. In 1938, this unit was to be utilized during the Sudetenland crisis. Additionally, KGrzbV 2, 4, 5 and 6 were combined, but they were disbanded shortly afterward.

When the Poland Campaign began, only KGzbV 1 and 2 existed and were under the direct command of the Luftwaffe Chief of Staff. They were only deployed on September 25 1939 for the bombing of Warsaw. By that time, combat bombers had already been withdrawn from Poland and Stukas could not drop incendiary bombs, so the bombs were rolled out of the loading hatches of the Ju 52's.

The first big deployment for the transport units, which were equipped almost exclusively with Ju 52's, came during the "Weserübung" on April 9th 1940. For the assault, aircraft staged out of Oslo, Stavanger, Aalborg-West and Aalborg-East. Deployed were KGzbV 1 and KGrzbVV 101 to 107. Additionally, five Lufthansa Ju 90's two Fw 200 and an old Junkers G38 were utilized. The air operations in the Denmark and Norway region were conducted under the X. Fliegerkorps, Generalleutnant Geisler commanding. Air assaults were directed by Oberstleutnant der Reserve Carl-August Freiherr von Gablenz, the Lufthansa technical director. Thanks to his assaults, the Oslo-Fornebu airfield was not blockaded. The Norwegians had pushed automobiles onto the runway and had set some of them on fire in order to hinder their landing. When, in reacting to the initial misleading reports, GenLt. Geisler ordered the takeoff of all Ju 52's remaining in Germany, in Fornebu, 50 Ju 52's had to find

Generalmajor Wolfram v. Richthofen as commander of the "Legion Condor" in 1938 in Spain.

Moroccan troops in Tetuan await transport to Spain.

The first bomber squadron of the Franco Air Force also flew the Ju 52-3m.

A Ju 52-3m ge of the Franco Air Force conducting operations.

The "moving truck" marking reveals the use of this aircraft as a tranporter.

A Ju 52-3m g3e of another squadron of the Franco Air Force.

places to land every 150 minutes next to the arriving assault forces. A transportation command (land) was created in Oslo under Major Beckmann, which was responsible for the lines of transport to Drontheim. We will see Beckmann again during the Demyansk operation.

A special chapter of the "Weserübung" was Narvik. The deployments were flown partly out of Oslo, partly from Hamburg-Fuhlsbüttel. On 13 April 1940, a particularly risky deployment was to be flown. It was to be a flight without the possibility of a return. Thirteen Ju 52's of KGrzvV 102 under the leadership of Oberst Baur de Betaz were to bring a fully-equipped mountain artillery battery to a frozen lake 15 kilometers north of Narvik.

Taking off from there again was not very likely. Some of the aircraft were taken from the Flugzeugführer-B-Schule pilot school in Winer Neustadt. After an intermediate landing in Oslo, the first aircraft landed on the Harstvigvaansee. Two aircraft nosed over during the landing, the others landed normally. Three aircraft lost their bearings and landed in Sweden. They were Ju 52's SE+HU, SE+KC and SE+IM. They returned to Germany in the beginning of September with swedish markings SE-AKR, SE-AKS and SE-AKT. The other aircraft remained on the frozen lake and were fired upon and later sank when the spring thaw came. One of these aircraft was able to be recovered at the end of August 1986, three others are to follow. The restoration of the first aircraft is already complete. Only KGrzbV 106 and 107 stayed in Norway and were assigned to Luftflotte 5.

The next test for the Ju 52 came on May 10th 1940 with the start of the German offensive in the west. Deployed were KGzbV 1 under Oberst Morzik and 2 under Oberst Conrad, who also commanded KGrzvB 9, 11, 12 and 172. All together, 430 Ju 52 were to be deployed, but in actuality is was only 401. Those units assigned under Oberst Conrad were to place the 22nd Luftlande-Infanterie-Division on airfields in the vicinity of The Hague. KGzbV 1 was to transport the 7th Flieger-Division with 4,000 men to their deployment areas. A total of 18,500 men were to be transported by air. A special mission was assigned to the "Sturmgruppe Koch," which was created out of the Gruppen "Granit" (Oblt Witzig), "Beton" (Lt Schlacht), "Stahl" under Oberleutnant Altmann and "Eisen" under Leutnant Schächter.

This Ju 52-3m g4e belonged to Kampfgruppe zbV. 106 and is seen here over Norway.

Another Ju 52, probably from a flight training school, seen on a Norweigan airfield after landing.

Operation "Weserübung:" arrival of Generaloberst Stumpf in Kyevik on April 22nd 1940.

"Weserübung:" Ju 52-3m g4e of KGrzbV 105 en route to Norway.

"Weserübung:" Oslo-Fornebu in early April 1940 seen with occupying German aircraft.

"Weserübung" – Ju 52-3m g4e of KGrzbV 106 at Oslo-Fornebu.

These units were to utilize 40 Ju 52's to tow DFS 230 gliders behind enemy lines and occupy Fort Eben Emael, the bridges over the Albert canal near Vroenhoven, Veldwezelt and Kanne and wait for the army units to arrive. For the most part, these missions were successful, despite a couple of failures during the support of the engineer Bataillons 51 under Oberstleutnant Mikosch, the Flakbataillon Aldinger with its 8.8cm Flak 36 anti-aircraft guns, the Henschel Hs 123 of II./LG 2 and the Stukas of StG 2. In comparison, the air-assault operations in Holland were sacrificial for the Ju 52. Landings on the Katwijk, Jijduin and Ypenburg went wrong. Of the units KGzbV 11 and 12 belonging to I./KG 172 only a few aircraft returned. Most of them fell to the bitter Dutch air defense. KGrzbV 1 lost over 60 aircraft, KGrzvB 2 about 140 to 150. With a total of 430 Ju 52's deployed, that means a loss of 51%! The Dutch were able to recapture the airfield in the area of The Hague, but they were not able to destroy those Ju 52' which had landed in the area of Valkenburg and Ockenburg. The Dutch commander-in-chief General Winkelmann signed the capitulation of all Dutch armed forces on 15 May 1940. The value of the cargo gliders had been proven, the German armor offensive in the Ardennes had succeeded, but almost half of the German air transport capability had been destroyed. About 100 Ju 52's were able to be taken back and repaired. General Walter Speidel commented: "The loss of air transport will effect us for years to come." Kampfgruppen zbV 11 and 12, because they had been almost totally wiped out, were dissolved.

A crash-landed Ju 52-3m at Oslo-Fornebu, June 22 1940.

Below: Dropping paratroops from Ju 52-3m g3e's of Kampfgeschwader zbV1.

Western Campaign, 1940: a Ju 52-3m g3e towing a DFS 230 cargo glider.

Below: Western Campaign: Paratroops during the jump.

After the capitulation of France a treaty was signed on July 23, 1940, with the Vichy government according to which the French aircraft manufacturers were to build 2,000 planes for the German Luftwaffe. The Amiot firm in Colombes was designated as the company to build Ju 52's. Operation Sea Lion seemed to be the next opportunity for the deployment of Ju 52's on a large scale. In September, the KGzvB 1 and 2 units were transferred to France. The operation did not come to be, however, and the air transport units returned to their homeland.

In the mean time, Mussolini had begun offensive operations in Libya against Egypt and against Greece in Albania, which both proved to be mistakes. On 18 December 1940, Hitler signed "Directive number 21" which ordered the invasion of the USSR. Prior to this, Operation Marita, which was to support the beaten Italians in Greece, was to be carried out. On 9 December 1940. 17 Ju 52's of III./KGzbV 1 transferred to Foggia, in order to provide support to the Italians in Albania. On 9 January 1941, Hitler decided that, due to the terrible defeats of the Italians in North Africa, he would also start an offensive there. The III./KGzbV 1 had been strengthened to 53 Ju 52's. The transported 1,665 italian soldiers to deployment areas, and an additional 2,905 tons of materiel to Albania and 10,970 men back to Foggia, among which were 8,730 wounded. On 6 February 1941 Operation Sunflower began, which was the deployment of German units to North Africa.

This Ju 52 of KGzbV 1 lies shot up near The Hague.

A crash-landed Ju 52-3m near Ypenburg.

Shot-up Ju 52's of 11./KGzbV 1 are arranged in rows along the highway between Rijswijk/Delft and Ypenburg.

Dutch soldiers with a shot-down Ju 52.

A Ju 52-3m of KGrzbV 172 during the Western Campaign.

Hitler's timetable was thrown off by the attempt on his life. On 25 March 1941 he issued Directive number 25: "Yugoslavia, its government and its military, is to be destroyed." At the beginning of February, KGrzbV 40, 50, 60, 101, 104 and 105 were taken out of the training units of the Chef des Ausbildungswesens der Luftwaffe, General Kühl, and placed under the Chef des Fliegerkorps. At the beginning of April, elements of the 1st and 2nd KGzbV brought the 22nd Luftlande-Division from Vienna to the Romanian oil fields. KGrzvB 104 was transferred to Athens-Tatoi. The II./KGrzbV 1 flew supplies from to Tripoli and Bengazi from March to April, turning this mission over to III./KGrzbV 1, which was stationed at Comiso on Sicily. On April 26th, at the end of the Balkan campaign, the air assault operation against the Isthmus of Corinth began, which was attended by KGrzbV 2, I./LLG 1, I. and II./KgrzbV 1, KGrzbV 60 and 102 and a squadron of Ju 52's with DFS 230 gliders. The men of the transport glider units did not succeed in securing the bridge across the Canal of Corinth, but at the same time, the airborne engineers were able to build a makeshift bridge next to that one which had been destroyed. On April 30th, the Balkan campaign ended.

A large portion of the British forces evacuated from Greece were brought to Crete, which was to become the next target of a German air assault operation, Operation Merkur.

The German side had made great underestimations about the strength of the forces found on Crete. Only about 20,000 men were expected to be there, but in fact there were over 42,640! Generaloberst Löhr, commander of Luftflotte 4 Südost was in charge of the Merkur operation. The landing itself was conducted under the command of General Student and his XI. Fliegerkorps. Air support was to be provided by the VIII. Fliegerkorps under General Wolfram von Richthofen. About 500 Ju 52's were ready to perform transport and glider towing duties. Gliders numbered 85, most of which were DFS 230's. Transport groups involved were I. and II./KGzbV 1, I./LLG 1 (LLG = Luftlandegruppe), as well as KGrzbV 40, 50, 101, 102, 105, 106 and 1./kgrzbV 172. Staging airfields were Corinth, Megara, Tanagra, Topolis, Dadion, Phaleron and Eleusis. The paratroop assault regiment arrived between 1 and 10 May 1941 as did the rest of the 7. Fliegerdivision; the 5. Gebirgsdivision and portions of the 6th were assigned to General Student.

A crash-landed Ju 52-3m in 1940 near Hirson in France.

Ju 52-3m g4e, G6+LF, of KGzbV 2.

Ju 52-3m g4e, NH+NZ, of KGrzbV 106.

Crete, December 1941: a Ju 52-3m of KGrzbV 400.

It was not until after very heavy and costly fighting that the situation on Crete stabilized in favor of the Germans. On May 27th the British troops began to evacuate, which was finished by the 31st. The losses were very heavy on both sides. But the worst of it was another blood letting for the air transport units. 151 Ju 52's lay wrecked on the island of Crete — victims of anti-aircraft and artillery. One had to realize that most of these were aircraft taken from flight schools, and would now be unavailable for training. On June 22 1941, Operation Barbarossa, the attack on the USSR, began. The following Ju 52 units were readied for participation:

Luftflotte 5, Scandinavia   KGrzbV 108
Luftflotte 1, Northern Sector KGrzbV 106
Luftflotte 2, Central Sector   KGrzbV 102
Luftflotte 4, Southeast KGrzbV 50 and 54

This meant about 200 transport aircraft along a front stretching 6,000 kilometers. One must realize that no existing transport units were involved, but rather elements surrendered from training units whose crews often did not even know one another. The fast-paced advance of the German armor elements and the rate of consumption of fuel, ammunition and provisions presented the transport units with almost impossible missions, especially as aircraft losses to anti-aircraft operations were considerable.

On October 1 1941, Oberst Gablenz, until now the commander of the air transport services, was promoted to General Chef des Planungsamtes im RLM. His replacement was Oberst Morzik. Aside from KGzbV 1, all Ju 52 units were now on the Eastern Front. And then came the first snowfall! On November 18 1941 an British counteroffensive drove Rommel's troops so far back that the fortification at Tobruk had to be abandoned. Naturally, this placed an increased demand on air transport. New air transportation units had to be formed out of school units once again, and these were KGrzbV 300, 400 and 500. Because the cold season in the East practically kept all supplies on the ground, KGrzbV 600, 700, 800, 800 and 999 were all formed out of the same arena. KGzbV 400 and 500 were created in half normal strength and transferred to he Mediterranean, Foggia, Brindisi and Trapani. In addition to these two units, III./KGzbV 1, the transport squadron of the Afrikakorps and the transport squadron of the X. Fliegerkorps were also stationed at these locations.

At this point in time is when two new members of the Ju 52 family appeared: the "Mausis" and the Ju 252.

The "Mausis" owes its creation to an idea of the Regierungsrats of the time (and later Ministerialdirigent) Th. Benecke. After studying a captured British magnetic mine, he suggested equipping a Ju 52 with a horizontally mounted spool 14 meters in diameter and winding aluminum cable about it 44 times. Electricity would be fed to this Ward-Leonard unit by a headlight battery. A Ju 52/MS thusly equipped flew over the harbor entrance at Vlissingen, where the British were constantly conducting aerial mine-laying, at an altitude of ten to twenty meters. Because the mines were equipped with delayed fuses of seven seconds, they detonated about 2-300 meters behind the aircraft. Benecke received a special award for this idea from the Generalluftzeugmeister. By 1940, the Sonderkommando Mausi special unit had been created, from which the 1. Minengruppe was later created. They were stationed at Jever harbor. Other MS-units followed, which were stationed on the West Friesland islands in Grossenbrode, Warnemünde and Cammin when the British dropped mines into the Baltic. The chief area of operations for the "Mausis" was the French Atlantic coast. After Operation Barbarossa had begun, there were always requirements for new "Mausi" units. They were operational from the Atlantic to the Black Sea.

At the end of October 1941, the Ju 252 V1 made its maiden flight. It had a long history of development behind it. In 1938 the Junkers

Markings of a Ju 52-3m belonging to IV./KGFzbV 1 in Africa 1941.

Original personal markings of one of the pilots of 2./Luftlande-Geschwader 2 on the engine cowling of a Ju 52.

people began to consider making improvements to the Ju 52, and the EF 77 plan came into being. This, too, was of three-engine construction, but without the corrugated skin characteristic of the Ju 52. The design was turned down by the RLM, however. But in July of 1939, the design was embraced once more, when Lufthansa made their desires for just such and airplane known. The outbreak of war disrupted the development of the plane. It was not until 1940, when it was widely believed after the victories on the

Ju 52-3m's flying supplies to the Afrika-Korps.

Collision between two Ju 52' during the supply flights to Africa.

Western Front that the war had been won did Junkers receive the directive to continue work on the aircraft's development. But now, instead of the normal 21 passenger load, Lufthansa required enough room for 32 to 35 passengers.

And so, the development of an entirely new aircraft was called for, which was approved in the spring of 1940 and a contract was given for three prototype aircraft. The most important feature of this new Ju 252 aircraft was the trapeze hatch in the floor, which had already been tested in the Ju 90 V 5 and V 6. Testing on the Ju 252 V 1 continued until Spring of 1942. At that time, the author had the opportunity to view the Ju 252 V 3 in Dessau. Using the trap door, it was easy to climb up into the fuselage from directly beneath the aircraft. A comparison of the interior with a nearby Ju 290 showed the cargo area of the Ju 252 did not lack much in comparison to the Ju 290. The aircraft, equipped with three Jumo 211 engines also had very good flight characteristics. While the Ju 252 V 1 through V 3 models went unarmed, the Ju 252 V 4 was fitted with weapons on the upper fuselage in the form of an electrically operated EDL 131 machine gun. The Ju 252 V 3 and 4 had, by this time, been deployed to North Africa for troop testing.

On January 3 1942, a portion of the 216th Infanterie Division, about 4,000 men, was encircled near Suchinitchi by the 10th Soviet Army. This was the first time an isolated unit had to be supplied completely from the air until it could be relieved on January 24th. About 40 Ju 52's were sufficient for this task. On 9 January 1942, Cholm was encircled. However, a much larger catastrophe loomed on the horizon near Demyansk when the II. deutsche Armeekorps, under the command of General von Brockdorff-Ahlefeld, and a large part of the X. Armeekorps was encircled on January 18th by the 11th and 34 Soviet Armies. On February 18, air transport commander Oberst Morzik was ordered to transfer General Oberst Keller and his combat staff to Luftflotte 1. It would take a minimum of 300 tons per day to supply the cauldron of Demyansk. Every available transporter which could possibly be used to carry supplies to Demyansk was made ready; KGrzbV 9 and 172, 600, 700, 800 and 900, as well as IV./KGzbV 1. Luftflotte IV had to give up its II./KGzbV 1 and KGrzbV 500 for the deployment to Demyansk, as well as the "Posen" and "Oels" KGrzbV's and portions of KGrzbV 105. In the beginning of March, KGrzbV 4, 5, 6, 7 and 8 were called up from training elements. Aside from KGrzbV 5, which was equipped with Heinkel He 111's, all participating units flew the Ju 53-3m. Staging airfields for the airlift to Demyansk were Plescow-West, Plescow-South, Korovyu-Selov, Riga, Riga-North, Dünaburg and Tuleblya. Because Oberst Morzik had only 75 Ju 52's available to him in the initial weeks, Armee-Oberkommando (AOK) released its KG4, which, with its He 111 H's, could only drop supply containers from the air. The He 111's also flew supplies to the Cholm bridgehead, although no landings were possible.

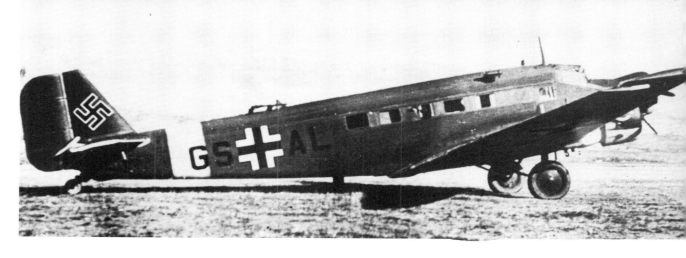

Eastern Campaign 1941: Ju 52-3m g4e, GS+AL in Bulgaria.

The Ju 52 was also used for civilian travel. This Ju 52-3m, WL-ABUX, is at the Hungarian Buda-Örs airport.

November 2 1941: Ju 52-3m, P4+HH, of Transportstaffel Nord(Ost) in Björnborg (Finnland) prior to taking off for Oslo.

Below: Ju 52-3m, serial number 5348, 7U+DM of KGrzbV 107, from which the Transportgruppe 20 was later created, in Norway.

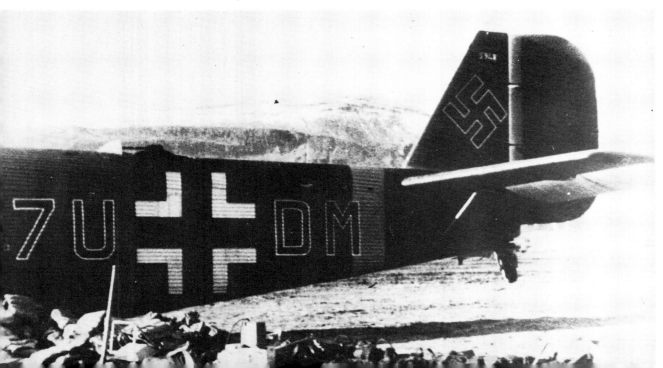

The first 40 Ju 52's landed in Demyansk on February 20 1942. Demyansk had to be supplied from the air for three months, which ended when a ground link could be established. Never before had this kind of performance been achieved by transport aircraft: in 33,086 sorties conducted from 20 February to 21 April 1942, 64,844 tons of materiel, 30,500 soldiers were flown into Demyansk and 35,400 wounded were taken on the return leg.

The air transport services lost 387 flyers, which equates to roughly 100 crews. The supply success of Demyansk had another consequence, and that was to convince those involved that it could be done again when the 6. Armee became encircled at Stalingrad, which was simply not possible due to the nature of this particular situation.

On November 19 1942 the two Soviet army groups, the Southwest Front and the Don front met up for a major assault, broke through the Romanian 3rd Army front and met up with the 3rd Army Group "Stalingrad Front" to form a pincer attack which enclosed the 6. Armee, the IV. Armeekorps, the 20th Romanian Infantry and the 1st Romanian Cavalry Division. Who bore the responsibility for supplying Stalingrad from the air is debatable. According to testimony from the Chiefs of Staff of the Heere and the Luftwaffe, Göring gave Hitler his pledge to supply Stalingrad from the air. The man in charge of supplying Stalingrad from the air was General Fiebig. These units participated in the operation: KGzbV 1 staff with KGrzbV 50, 102, 172, 900, all equipped with Ju 52's, KGrzbV 5, KG·27, KG 55, III./KG 4 with He 111 H's, and beginning in December 1942 I. and II./KGzbV 1, KGrzbV 9, 105, 500, 700, 900 with Ju 52's and 14 other groups with various types of aircraft.

Additionally, all available Ju 90, 290 and Fw 200's were sent to Stalino, including Hitler's Fw-200 used for personal travel as well as his pilot Hans Baur. The losses among the transport fliers grew from day to day. On January 18 1943 alone 30 Ju 52's were lost, 10 completely. If the soviet fighters had been utilized better employed, the losses would have been even higher. From November 25 1942 to January 11 1943, 5,227 tons of provisions were flown to Stalingrad. After that, Pitomnik airfield fell to the Russians so that supplies could only be dropped from the air. By December 24 1942 the Tazinskaya airfield, from which supplies were being staged, was within range of soviet artillery. General Fiebig wanted to evacuate the airfield, but Göring ordered all aircraft to remain on the ground! When Fiebig assumed responsibility and had the airfield evacuated, 109 Ju 52's managed to get away, 55 were destroyed on the ground. The Ju 52 units lost approximately 250 crews, both new and experienced, at Stalingrad – losses from which the transportation units would not recover.

The war had also become hopeless in North Africa. Rommel's last attempt at an offensive against the British 8th Army failed. The allied landing in Algeria and Tunis in the rear area of the German/Italian forces created a hopeless situation. The air transportation services suffered heavy losses here as well while transporting goods to North Africa. On April 5, 1943, US fighters engaged a Ju 52 unit, which had to fly without fighter cover, and shot down 14 aircraft. Ten Ju 52's were destroyed and another 65 were severely damaged on the ground at airfields in Sicily. On April 18 1943 24 out of 65 Ju 52's were shot down, another 35 were damaged, in spite of their fighter cover. On May 13,

A former Lufthansa Ju 52-3m g3e, April 1942 in Copenhagen/Castrup.

Below: A Ju 52-3m g6e air ambulance.

Engine maintenance on a Ju 52-3m g3e, which had been turned over to the Luftwaffe by Lufthansa.

Below: Landing gear damage on a Ju 52-3m g7e, CO+AL.

1943, the tragedy in North Africa came to an end.

Transport aircraft, the Ju 52's and Me 323's, had flown 8,388 soldiers and 5,040 tons of supplies to North Africa. Seventy-six Ju 52's and 14 Me 323 were shot down, 275 flying personnel had died or been wounded.

After these enormous losses, the transportation units were reorganized (May 15 1943). They were combined into the XIV. Fliegerkorps. The Korps headquarters was located in Tutow in Pommern. Ju-52-equipped Transportgeschwader's (TG) 1 through 4 were created from KGzbV and KGrzbV, while TG 5 was equipped with Me 323's.

Supplying the 17. Armee at the Kuban Bridgehead by air the air began on February 4 1943 and continued until their evacuation in September – October 1943. An average of 182 tons were transported daily using from 40 to 45 Ju 52's.

Beginning at the end of October 1942, further development of the Ju 252 was being conducted at Dessau, no longer as an all-metal aircraft but a combination of materials due to lack of raw materials. The Jumo 211 engines with which the Ju 252 had been equipped, was also suffering production bottlenecks, so the new aircraft, now called the Ju 352, had to be equipped with the BMW-Bramo engine which was readily available. Because many portions of the Junkers factory were no longer operational due to bombing, the construction of th Ju 352 was transferred to the facilities at Fritzlar airbase. This is where the Ju 352 conducted its maiden flight on October 1st 1943. This aircraft, and the Ju 352 V 2 which followed were not armed. The Ju 352 A-1, which was mass-produced in limited numbers, had one machine-gun position for self-defense on the top

Ju 52-3m g5e's of KGrzbV 106.

Eastern Front, winter of 1941: Ju 52-3m of KGrzbV 105.

Wounded were taken along on return flights from the Front.

Supplying Demyansk, Spring 1942: Ju 52-3m's of KGrzbV 500 in Teleblja prior to flying to Demyansk.

Eastern Front, 1941/42: Ju 52-3m G6+AF and G6+DY of KGrzbV 106.

portion of the fuselage right behind the flight deck, with an MG 151/20 machine gun mounted in a plexiglas cupola. The Ju 352 was the first German aircraft to have completely reversible propellers, which were constructed by the Vereinigte Deutsche Metallwerke (VDM) in accordance with a design from the Messerschmitt firm. Those few Ju 352's produced prior to the war's end were tested on the front lines. One Ju 352 was flown as a test aircraft in Czechoslovakia after 1945. The author saw the last Ju 252 on the edge of Staaken airfield near Berlin in the Fall of 1945, where it was sitting together with many other German aircraft on the Berlin-Hamburg rail line.

The Soviet attack on German Heeresgruppe Nord and the northern flank of Heeresgruppe Süd began on January 27 1944, and at approximately the same time, the U.S. Army landed in Italy near Anzio and Netuno. In the east, portions of the German army were time and again encircled by the Soviets. Again and again the Ju 52's were called upon to supply those encircled units. In January and February 1944, the II. and III./TG 3 together with I./TG 1 supplied the 8. Armee which was encircled near Tscherkassy. The supplying of the 17. Armee, which was cut off in the Crimea, had begun in November 1944 and continued until the final withdrawal of German troops in May 1944. The 1. Panzerarmee which was cut off in the area of Kamenez-Podolsk von Lemberg, was supplied by air beginning in the middle of May 1944. In may 1944 there were only about 550 to 600 Ju 52's on hand, but not all were deployable. After the invasion on June 6th 1944, the supply of Ju 52's from the Amiot works in Colombes near Paris stopped. This is where the last Ju 52's were built. In February 1945 the General Staff of the Luftwaffe ordered extra fuel be supplied

Eastern Front, 1941/42: Ju 52-3m g4e belonging to a courier squadron.

Below: Nosed-in during landing: Ju 52-3m g6e, 1Z+F of Kampfgeschwader zbV 1.

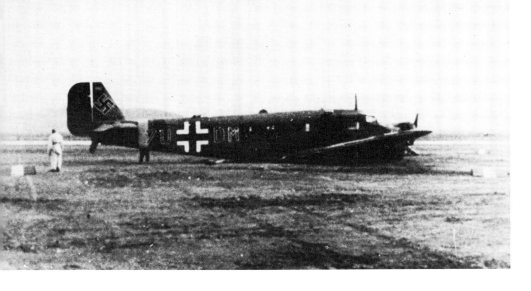
A belly-landing for a Ju 52-3m, 7U+DM, belonging to KGrzbV 108.

This Ju 52-3m belly-landed in the summer of 1941 near Chiesch (USSR).

Ground support personnel unloading ammunition cases from a Ju 52.

Changing engines at a front-line airfield.

This Ju 52 of Flugzeugführerschule (B) 8 in Wiener Neustadt returned after flying to the front on a mission for KGrzbV and had to make an emergency landing after tangling with a fighter. The propellor is mounted temporarily.

Norway, 1941: A Ju 52-3m of KGrzbV 108 in Banak.

The Romanian allies also recieved a few Ju 52-3m's. Here, fuel tanks are being unloaded.

This Ju 52-3m, NO+IJ, serial number 2977/13330, became lost during a courier flight and crash-landed on February 24 1943 in Lekvattnet, Sweden.

Ju 52-3m g3e, a former Lufthansa aircraft, of Transportstaffel Nord (Ost) in 1941 in Rovaniemi, Finland.

Below: A Ju 52-3m with the anti-mine ring called "Mausi."

to the Ju-52 mine sweeper units due to the increased danger of mines in Belt and in the Baltic. December 1945 marked the beginning of the air supply to Budapest, which fell at in February 1945. The most costly operation came in the airlift to Breslau beginning in February 1945 during which 165 transport aircraft, mostly Ju 52's, were lost along with their crews. Breslau capitulated on May 6 1945. Similar operations continued to be successful until war's end in such areas as Petrikau, Posen, Schneidenmühl, Arnsewalde, and Glogau. Ju 52's also participated in the Ardennes offensive in December, when 67 Ju 52's took part. The last reliable reports of on-hand Ju 52's came on 25 April 1945:

I./TG 1 in Windau/Kurland: 11 Ju 52's
Tutow staff: 14 Ju 52's
II./TG 1,3. Staffel in Pütnitz: 15 Ju 52's
II./TG 2 in Shrasslavitz: 31 Ju 52's
III./TG: 2 in Klattau
I./TG 3 in Neuenburg: 24 Ju 52's
II./TG 3 in Güstrow: 34 Ju 52's
III./TG 3 in Wimsbach: 37 Ju 52's
Tr.Gr. 20 in Oslo/Norway: 38 Ju 52's
See.Tr.Staffel 2 in Hommeloik/Norway: 7 Ju 52's

There were an additional 23 Ju 352's in Tutow.

All together, 190 Ju 52's and 23 Ju 352's were combat-ready. On April 20th 1945, the last Lufthansa Ju 52-3m took off from Berlin-Tempelhof airport, made one layover stop at Munich-Riem airport, and then disappeared. Its fate has never been discovered!

This, however, was not the end of the Ju 52's history. It was decided in France to continue construction of the Ju 52-3m g10e as the A.A.C.1. (A.A.C. meaning Ateliers Aeronautiques de Colombes). Eight Ju 52's formed the basis of the new "Air France." Over 400 Ju 52's were built in Colombes, 216 of which went to the French air force which was deployed to the frontlines in Vietnam from 1947 to 1952. Thirteen of these Ju 52's went to the Portuguese air force. One A.A.C.1 is in the Deutsches Museum in Munich. The Ju 52-3m also formed the basis of the Spanish air force. The Ju 52 was used as the basis for the construction of the CASA (Construcciones Aeronauticas S.A.) 352 of which 170 were built. Of those, one was still in service with the 361st Escuadron in 1970-1971 until it was replaced by the Douglas C-47. Three Ju 52's formed the core of the "Rijksluchtvaartschool." Five Ju 52's were being utilized as training aircraft by the Belgian air force as late as 1950. In Switzerland, three Ju 52-3m g4e's have been kept airworthy despite have been retired due to the loving care of the swiss people. They still make occasional sightseeing flights.

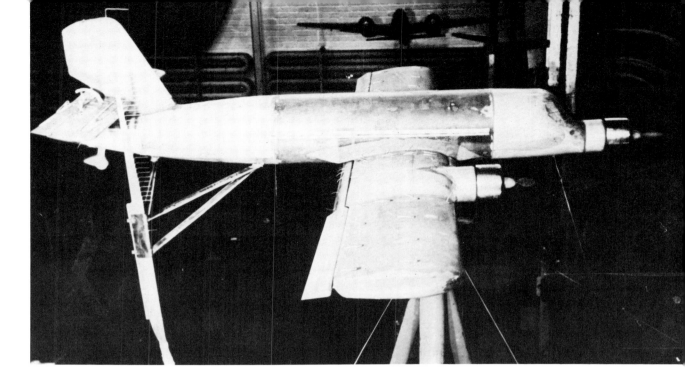

Above: Wind tunnel model of the EF 72, original form of the Ju 252.

Right: Ju 252 V1 in the final assembly hall at Dessau.

Maiden flight of the Ju 252 V1 with landing gear extended.

Ju 252 flight deck with the pilot and co-pilot.

Ju 252 V4, T9+AB, at the testing unit of the Oberkommando der Luftwaffe (OKL).

Ju 252 V 5, DF+BQ, of Lufttransportstaffel (LTS) 290 in 1943 in the southern sector of the Eastern Front.

Ju 252, V5, DF+BQ, of Lufttransportstaffel (LTS) 290 in the summer of 1943 in Brasov, Romania.

The German Lufthansa airline has succeeded in restoring a Ju 52-3m to airworthy status in spite of great difficulties. The most spectacular rescue of an old Ju 52 is currently in progress in Norway: a German group call the "Interessengemeinschaft Ju 52 e.V." is attempting to recover one of a dozen German Ju 52's which landed on a frozen lake during operation "Weserübung" and later sank during the Spring thaw. Forty-six years later, they were able to hoist one of the aircraft in surprisingly good condition and take it back to Germany, where it will be restored at Lufttransportgeschwader 62 in Wuntsdorf. It is hoped that other aircraft can be recovered. These might well be the last German Ju 52's which perpetuate the memory of what is probably the most famous of German aircraft.

A Ju 352 V4 in Fritzlar.

This Ju 352 landed in Denmark in May 1945 due to lack of fuel.

A Ju 352 in Czechoslovakia after the war.

This Ju 52 of TG/50 remained in Kjevik in 1945 and was taken over by the Norwegians.

This Ju 52, DP+FJ, serial 640416, landed in Bonarp, Sweden on May 2nd 1945.

This "Mausi" version of a Ju 52-3m g63, serial number 3428, flew mine speeping operations after the armistice in 1945 with English registration VM 927 and then was transferred to the Dutch "Rijksluchtvaartschool."

A French Ju 52 built in Colombes as A.A.C. 1.

A Spanish-built C.A.S.A 352L.

Ju 52-3m, serial number 5715, of the dutch "Rijksluchtvaartschool" carrying registration PH-UBA "OPA."

One of three Ju 52-3m g4e's delivered to Switzerland over 50 years ago. It was saved by the swedish population from being scrapped and still conducts sightseeing flights today.

A Spanish CASA 352 L with German markings and american registration N 9012P belonging to the Confederate Air Force in Texas.

Although seemingly ready for the scrap heap, this "Iron Annie" which at one time carried serial number 5489, landed at Hamburg-Fuhlsbüttel for restoration on December 28 1984.

It flew again at the Hamburg International Airshow in 1986. Only the sound of the engines was different from that of the old Ju 52.

Sightseers and visitors to the Frankfurt airport can see this Spanish-produced version of the Ju 52-3m, this memorable German airplane.

This Ju 52 at the Frankfurt airport is an intriguing sight for all visitors who have not been able to escape their fascination with this plane.

Also from Schiffer Military/Aviation History

Also Available:
GERMAN FIGHTER ACE ERICH HARTMAN
STUKA PILOT HANS-ULRICH RUDEL
LUFTWAFFE RUDDER MARKINGS
THE LUFTWAFFE IN NORTH AFRICA